Four Firefighters Die
in Seattle Warehouse Fire
Seattle, Washington

Investigated by: J. Gordon Routley

This is Report 077 of the Major Fires Investigation Project conducted by TriData Corporation under contract EMW-94-C-4423 to the United States Fire Administration, Federal Emergency Management Agency.

Department of Homeland Security
United States Fire Administration
National Fire Data Center

U.S. Fire Administration Fire Investigations Program

The U.S. Fire Administration (USFA) develops reports on selected major fires throughout the country. The fires usually involve multiple deaths or a large loss of property. But the primary criterion for deciding to do a report is whether it will result in significant "lessons learned." In some cases these lessons bring to light new knowledge about fire--the effect of building construction or contents, human behavior in fire, etc. In other cases, the lessons are not new but are serious enough to highlight once again, with yet another fire tragedy report. In some cases, special reports are developed to discuss events, drills, or new technologies which are of interest to the fire service.

The reports are sent to fire magazines and are distributed at National and Regional fire meetings. The International Association of Fire Chiefs assists the USFA in disseminating the findings throughout the fire service. On a continuing basis the reports are available on request from the USFA; announcements of their availability are published widely in fire journals and newsletters.

This body of work provides detailed information on the nature of the fire problem for policymakers who must decide on allocations of resources between fire and other pressing problems, and within the fire service to improve codes and code enforcement, training, public fire education, building technology, and other related areas.

The Fire Administration, which has no regulatory authority, sends an experienced fire investigator into a community after a major incident only after having conferred with the local fire authorities to insure that the assistance and presence of the USFA would be supportive and would in no way interfere with any review of the incident they are themselves conducting. The intent is not to arrive during the event or even immediately after, but rather after the dust settles, so that a complete and objective review of all the important aspects of the incident can be made. Local authorities review the USFA's report while it is in draft. The USFA investigator or team is available to local authorities should they wish to request technical assistance for their own investigation.

This report and its recommendations were developed by USFA staff and by TriData Corporation, Arlington, Virginia, its staff and consultants, who are under contract to assist the Fire Administration in carrying out the Fire Reports Program.

The U.S. Fire Administration greatly appreciates the cooperation received from Fire Chief Claude Harris and many of the officers and firefighters of the Seattle Fire Department. Appreciation also goes to Assistant Chief of Operations David S. Campbell, Assistant Chief of Administration Gregory M. Dean, and Acting Fire Marshal Jerald A. Birt.

For additional copies of this report write to the U.S. Fire Administration, 16825 South Seton Avenue, Emmitsburg, Maryland 21727. The report is available on the U.S. Fire Administration's Web site at http://www.usfa.dhs.gov/

U.S. Fire Administration

Mission Statement

As an entity of the Department of Homeland Security, the mission of the USFA is to reduce life and economic losses due to fire and related emergencies, through leadership, advocacy, coordination, and support. We serve the Nation independently, in coordination with other Federal agencies, and in partnership with fire protection and emergency service communities. With a commitment to excellence, we provide public education, training, technology, and data initiatives.

TABLE OF CONTENTS

Four Firefighters Die in Seattle Warehouse Fire
January 1995

Local Contacts: Seattle Fire Department
301 2nd Avenue South
Seattle, Washington 98104
(206) 386-1400

Claude Harris, Chief of Department
David S. Campbell, Assistant Chief of Operations
Gregory M. Dean, Assistant Chief of Administration
Jerald A. Birt, Acting Fire Marshal

OVERVIEW

Four firefighters died when a floor collapsed without warning during a commercial building fire in downtown Seattle on January 5, 1995. The cause of the fire was determined to be arson, and a suspect was apprehended and charged with four counts of homicide.

The circumstances of this incident are similar to a number of other multiple fatality incidents that have claimed the lives of more than 20 firefighters in recent memory across the Nation.[1] The similarities include fires in buildings that have access at different levels from different sides, resulting in confusion over the levels where companies are operating. There have also been a number of situations where firefighters have been operating on a floor level that appeared to be safe, not knowing that they were directly above a serious basement or lower level fire. These incidents resulted in sudden and unanticipated floor collapses, which either dropped the firefighters into the fire area or exposed them to an eruption of fire from the level below them.

The structure involved in this incident was primarily constructed of heavy timber members. However, a modification to the structure had resulted in the main floor being supported by an unprotected wood frame "pony wall" along one side. The sudden failure of this support caused the main floor to collapse without warning. Firefighters working on this floor believed that they had already gained control of the situation and were not aware that the main body of fire was directly below them or that it was exposing a vulnerable element of the structural support system.

---

[1] Refer to previous reports in this series for additional information on incidents in Breckenridge, PA (Report 061) and Pittston, PA (Report 073). Twelve New York City firefighters died in a similar incident in 1966.

This situation illustrates that multiple firefighter fatalities can occur at an incident where fire suppression operations are well organized, well managed, and well executed with a strong emphasis on operational safety. One of the important lessons that come from this event should be valuable to both new and experienced fire officers; it shows how critical information can be missed during a complicated incident command operation, particularly when command officers are distracted by trying to perform too many functions without support staff.

SUMMARY OF KEY ISSUES

Issues	Comments
Unanticipated flaw in the structure.	The structural support that failed and caused the floor to collapse was out of character with the other structural components of the building and extremely vulnerable to a basement fire.
Unusual and complicated building configuration.	The building had been modified numerous times and was very difficult to "size-up" from exterior vantage points.
Companies entering from different sides did not realize that they were on different levels.	The attack crews entering from the east side did not realize there was a basement below them. The crews entering from the west side located the fire in the basement, but did not realize that the attack crews were above them.
No pre-fire plan.	The responding units did not have a pre-fire plan of the building. A plan could have helped them interpret the complex arrangement of the building and might have allowed them to recognize that the fire was below the main floor level.
Observed fire conditions presented a very different interpretation of the situation from different vantage points.	The approaching units observed a fire that appeared to be burning on the exterior and threatening to extend to the interior of the structure. Their attack plan was based on this interpretation of the fire.
Interior conditions were consistent with the initial interpretation of the fire.	Conditions encountered by crews making the interior attack were not inconsistent with their expectations and did not cause them to question the attack plan.
Operating personnel did not recognize the significance of their observations with respect to the overall incident.	Personnel who observed the fire from different vantage points had valuable information that could have caused the Incident Commander to revise the attack plan. However, they did not realize that their observations were new information that did not coincide with observations from other points.
Lack of progress reports.	The interior attack crews on the upper level did not report that very little fire had been found inside the building and all flames appeared to have been knocked down. The crews on the lower level did not report that they had found a large area that was fully involved in fire. The discrepancy between these reports would have alerted the Incident Commander to reevaluate the attack plan.
Although the first alarm companies had been warned that the building was a possible arson target, the warning did not provide them with information that would have changed the strategy or tactics.	The information about an arson threat had made the responding companies aware that the building was a potential arson target. They were not aware of the unusual interior arrangement or the vulnerable structural support.

DESCRIPTION OF STRUCTURE

The Mary Pang Chinese Food Company occupied a building in Seattle's International District at 815 Seventh Avenue South, a few blocks from the Kingdome. The company prepared frozen Chinese food dishes for distribution to grocery stores in the Seattle area and had been operating from the same location for more than 20 years. A bakery that supplied retail outlets in the Seattle area occupied part of the lower level, and an unused warehouse space was rented out as an evening practice area for a rock band.

The building was constructed in stages and had been modified several times over its history of more than 85 years. The occupancy is also believed to have changed several times. The resulting structure was very difficult to interpret from the exterior. The history of the building was obtained through a number of records and plans, dating back to 1909; however, many of the details are uncertain, particularly with respect to the dates and sequence of modifications.

The structure at the time of the fire was an L-shape, with two 2-story sections along the north-south axis and a single story wing extending to the west. The original section, which connects the north and west wings, is referred to as the center section in this report. The configuration is represented in Figures 1A and 1B on the following pages. The fire originated in the center section and spread to the north wing.

HISTORY

The oldest part of the Mary Pang building was constructed as a single story structure, 60 x 60 feet in area, with brick exterior walls and wood posts supporting a wood roof. In subsequent years two additional sections, also 60 x 60 feet in area, were added north and west of the original structure to form an L-shaped building. When the two sections were added, two of the brick walls of the original structure became interior fire walls. Over the years several doorways and windows were bricked-over and the connections between sections were modified several times.

When the original structure was built it was in a low-lying swampy area, south of the downtown business district. During the 1920's a major project was undertaken to raise the ground level in this part of Seattle. A large hill north of downtown was removed and the fill was used to raise the level of the low-lying area to the south. The streets in this area were raised by 10 to 20 feet and the ground floors of many buildings became basements.

At the corner of Seventh Avenue South and Charles Street the new grade level of the streets was higher than the roofs of the single story structures, which were now located in the low-lying area between the raised streets. The east wall of the original structure was reconstructed as a retaining wall and increased in height by approximately five feet to hold back the fill that was used to raise the level of Seventh Avenue South. The south wall of this section was partially buried, covering several of window openings. The original ground floor level became a windowless basement, accessible only through the north and west wings.

When the elevation of the streets was raised, the roof of the original 60 by 60 foot structure was removed and an upper story was added, with the new floor at the same level as the sidewalk on Seventh Avenue South. The new story was post and beam construction, with large dimension wood members supporting the new floor and roof. It was built with three wood frame exterior walls and a brick wall on the south side.

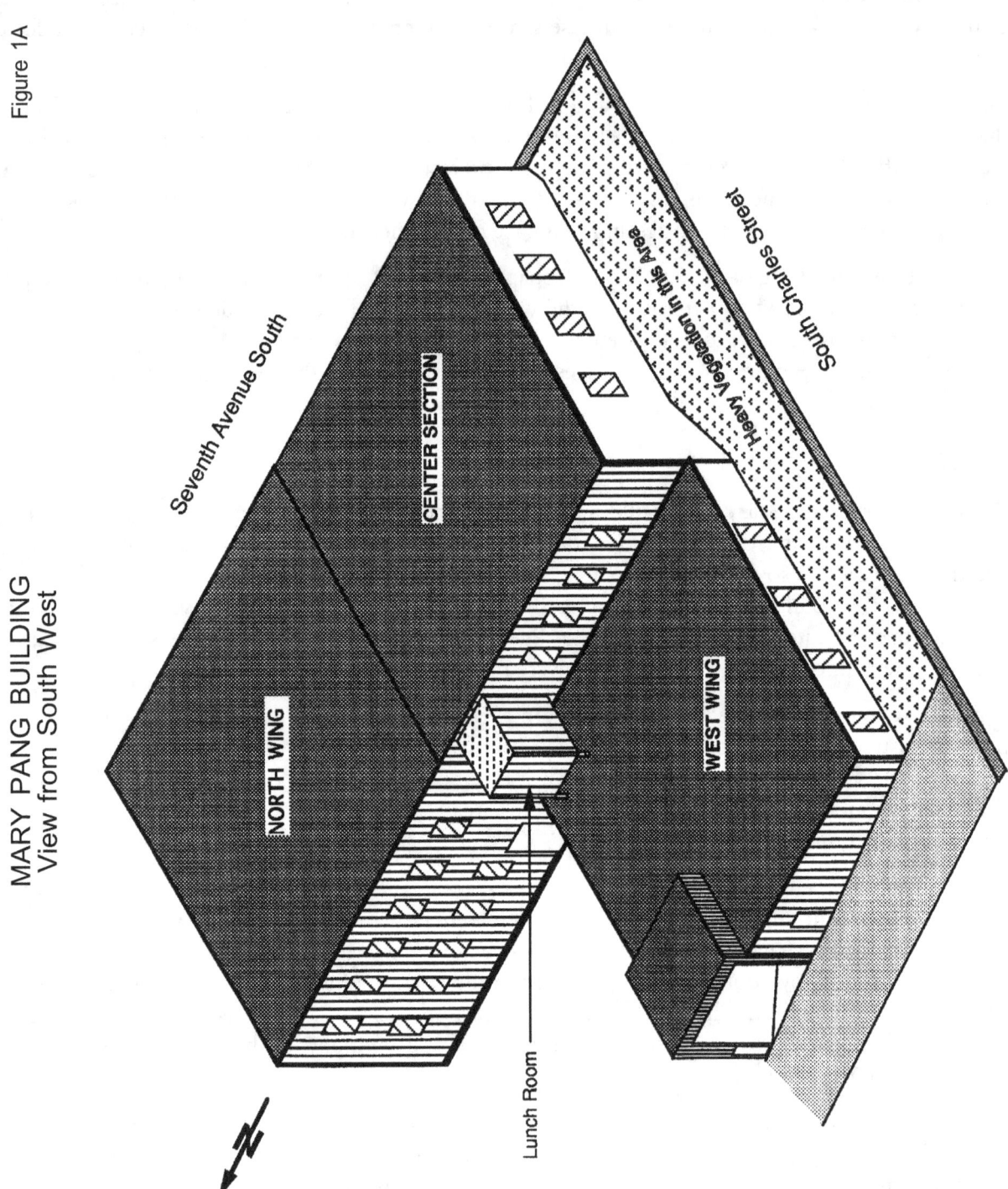

Figure 1A

MARY PANG BUILDING
View from South West

Figure 1 B

MARY PANG BUILDING
View from South East

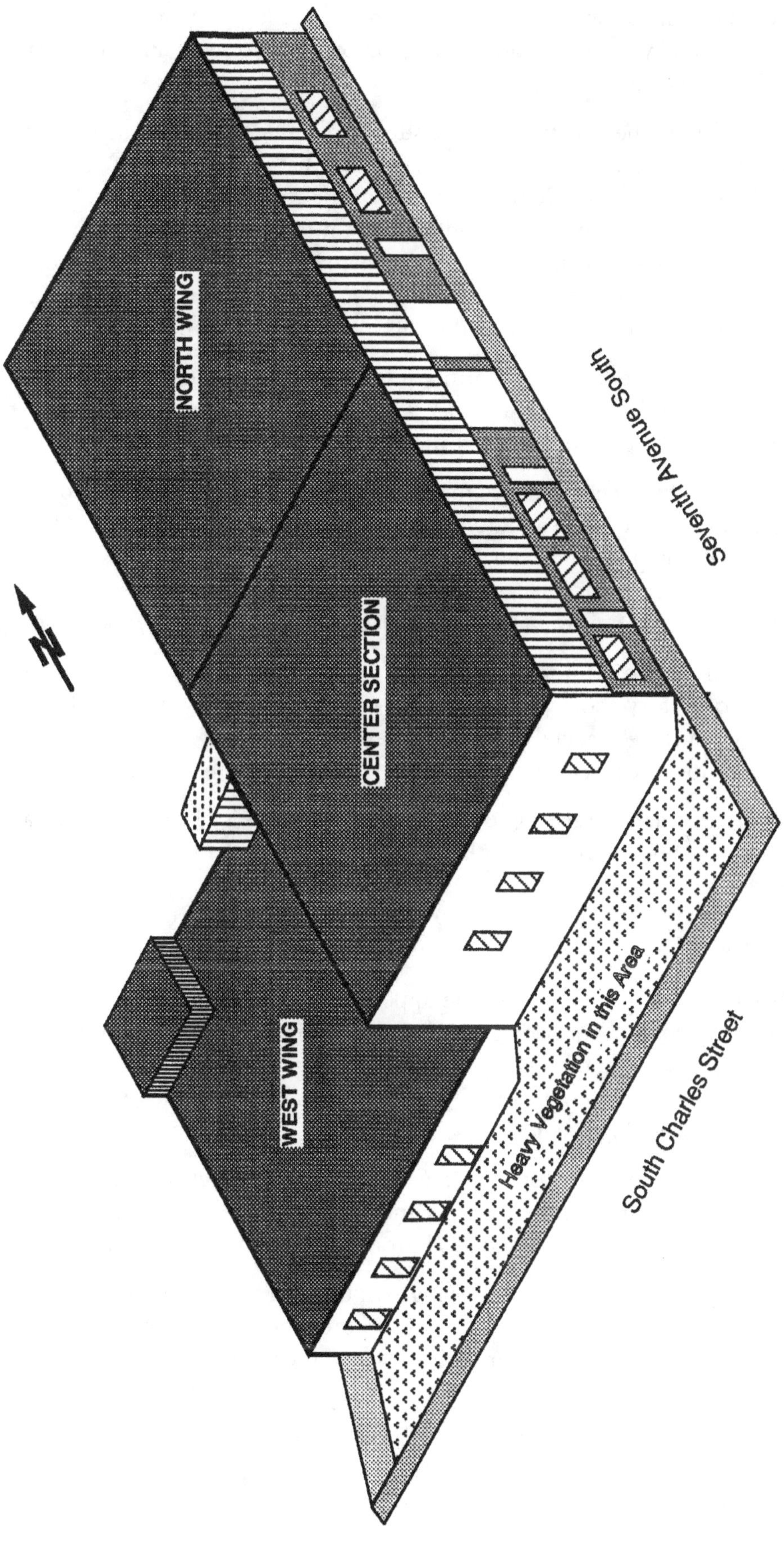

NORTH WING

CENTER SECTION

WEST WING

Seventh Avenue South

Heavy Vegetation in this Area

South Charles Street

The large mass of the wood structural members made them inherently fire resistant – they will burn and eventually collapse, but they can be expected to maintain their structural integrity for an extended period of time.

The hidden flaw in the structure was the support for the ends of the new floor joists along the north wall. The original roof was supported by a ledge that was incorporated in the brick wall. Because the new floor was four to five feet higher than the roof, a wood frame "pony wall" was fabricated from 2 by 4 inch wood members and erected on top of the ledge to support the ends of the new floor joists. This assembly did not have the inherent fire resistance of the more massive wood members and could be expected to fail rapidly under fire conditions. The failure of this assembly would release the ends of the floor joists and result in a sudden floor collapse.

An upper level was added to the north wing at a later date and the original brick wall between the sections was extended upward to create a fire separation between the upper stories. The original frame wall of the upper story was left in place.

Over the years several changes were made in the structure to meet the needs of different owners and occupants. This included the addition of a concrete topping over the wood floor to meet health standards for a food preparation facility. At the same time, the interior stairway was permanently covered-over with a wood and concrete deck, leaving no interior communication between the upper and lower levels. A small lunchroom was also added adjacent to the north wall, above the west wing.

At the time of the fire the north wing and the original (center) section, which faced Seventh Avenue South, had one story above grade level and one story below grade. The west wing remained as a single story structure attached to the lower level of the center section. The three sections were separated by interior brick walls; however, there were openings between the sections at each level.

Additional information relating to the construction details of the building can be found in a separate section of this report. Floor plans and diagrams of the structural details are also provided in that section.

OCCUPANCY

The Mary Pang Company occupied the upper level of the center section and all of the north wing of the building. Two doors provided access to the office area at 815 Seventh Avenue South. There were also two large roll-up doors on the Seventh Avenue side, one on each side of the interior brick wall. A single interior opening in this wall provided access between the two sections at this level and was equipped with a sliding fire door.

Most of the upper level was open area used for food preparation, packaging, and shipping, with a few small offices and rooms on the perimeters. There were several coolers for storing perishable goods and some large cooking and food processing equipment was installed in the open area.

Most of the lower level of the north wing was used for storage by the Mary Pang Company. Part of this space was rented out to the rock band to store their equipment and for evening practices. There was exterior access to this area through a door in the west wall near the point where the three sections connected.

Most of the west wing and half of the basement of the center section were rented by the bakery. A natural gas fired bake oven was installed in the basement of the center section. An interior loading dock and parking area, used primarily by the bakery, occupied the remainder of the west wing; several delivery trucks were parked in this area when the fire occurred. The bakery used only the rear access to the building.

The remainder of the basement in the center section was used for storage by the Mary Pang Company and the family that owned it. The access to this room was through a sliding fire door from the loading dock. A second access to this room was available through a narrow door in the north wing near the west wall. This door was seldom used because of its difficult access; to compensate for the difference in floor levels there were steep stairs on both sides of the narrow opening.

This storage room, which was 30 feet wide, 60 feet deep, and approximately 20 feet in height, was heavily loaded with combustible products.

There was no ceiling in the storage room. The underside of the floor deck and the structure that supported it were unprotected in this room, including the vulnerable "pony wall." The room had no windows, only two doors, and three brick walls. The partition between the storage room and the bakery was wood construction and there was a wood frame section along the upper part of the west wall, between the top of the brick wall and the underside of the floor deck. (Refer to the section on construction details.)

EXTERIOR APPEARANCE

The structure that could be seen from the exterior was very difficult to relate to the interior arrangement or the construction details. Because of the abrupt change in grade levels at the north side, it was impossible to drive and difficult to walk all the way around the building to make a complete "size-up."

The upper level was primarily wood frame construction; however, exterior architectural trim had been added to the entire east wall, which was the front face of the building. The trim included grouted stucco panels and decorative wood siding that gave the appearance of a single one-story structure and effectively disguised its age. There was no evidence of the interior brick wall or that there was a basement under this part of the building.

The south wall was obscured by heavy vegetation in the sloping setback area between the building and South Charles Street. This wall could only be closely examined from Seventh Avenue at the east side or from the driveway at the west side of the building. Toward the east end, the brick had been covered by stucco and there was a row of windows that coincided with the upper floor level. When viewed from the west, the age of the brickwork was evident and the ground floor windows were visible, just above the embankment. From either end the structure appeared to be one story; however, the parts that could be seen were two different stories.

The age and complexity of the building could be appreciated when viewed from the west side, which faced a driveway and parking lot. From this side the three different sections of the building were discernible. The wood frame exterior walls of the two story sections were covered by a shiny metal skin that gave the appearance of a warehouse or industrial building. Windows were visible on both levels of the north wing and above the west wing in the center section.

The west wing was a combination of brick and wood construction that appeared to be considerably older than the two story sections, because no cosmetic efforts had been made to conceal its age and condition. This wing appeared to have an unusually low roof line, which was caused by the slope of the driveway up to Charles Street; the floor inside was actually about four feet lower than the driveway. A steep interior ramp had been installed to provide vehicle access down to the loading dock, and the roof had been raised at the northwest corner to provide extra height for trucks using the ramp.

From the street to the north, when viewed from a distance, the building appeared to be a single story structure. Close to the building the ground dropped away to reveal that this section was actually two stories in height and covered by relatively new wood siding. The west wing was visible from this vantage point on the north side.

ARSON INFORMATION

Several weeks before the fire occurred arson investigators received information that an attempt would be made to burn this building during a specific time period. Surveillance was conducted for the period indicated by the informer and the first alarm companies were advised to beware of the potential for an arson fire at the location. The fire occurred 16 days after the surveillance was terminated.

The investigation determined that the fire was deliberately started in the basement storage room.

FIRE DISCOVERY

The fire was set in the storage room shortly before 1900 hours on Thursday evening, January 5, 1995. It quickly grew to major proportions within the storage room and burned through the wood frame partition into the area occupied by the bakery. It also burned through the metal clad wood frame section that covered the gap in the west wall, between the top of the brick and the underside of the upper floor deck. The fire then lapped up the exterior of the metal covered wall, above the roof of the west wing.

The first call to the Seattle Fire Department was received at 1902 hours from a member of the band, reporting that smoke was coming into their practice area from the adjoining section of the building. This was quickly followed by other callers reporting a working structure fire in the area of Maynard Avenue South and South Dearborn Street. A full first alarm assignment, consisting of five engine companies, two ladder companies, one BLS ambulance, one paramedic unit, an air supply unit, and two command officers was dispatched at 1903 hours.

Companies approaching the scene saw a large column of smoke in the air. The wind was calm and the smoke column was rising almost vertically above the fire. The first arriving company, Engine 10, responded to the intersection of Maynard and Dearborn, then proceeded south on Maynard to South Charles Street. From this vantage point a heavy volume of fire could be seen above the roof of the west wing of the building. The fire appeared to be burning against the exterior of the rear (west) wall of the center section and threatening to penetrate into the structure.

As Engine 10 passed to the south of the building along South Charles Street, some of the crew members noted that the main body of fire appeared to be coming from a small shed that extended out from the wall, above the roof of the west wing. (This was a lunchroom that had been added at the rear of the upper level.) They believed that the fire could be coming from this shed and extending to the main structure. Engine 10 proceeded around to the east side, which was the front of the building.

INITIAL ACTIONS

Engine 10 reported on the scene with a "well involved building fire" and assumed command at 1907 hours. The Lieutenant's report indicated that Engine 10 would be "laying a 1-1/2 inch manifold." Engine 10 hooked up to the hydrant at the northeast corner of Seventh Avenue South and South Charles Street and extended a 3-inch line to the manifold, which was placed at the southwest corner. Several attack lines can be taken from the manifold.

Ladder 1 followed Engine 10 into the scene and spotted on the east side of the building near the mid-point. The Lieutenant in charge of Ladder 1 conferred with the Lieutenant of Engine 10 while the crew initiated forcible entry. The crew was then directed to raise ground ladders and proceed to the roof to perform vertical ventilation.

The Acting Deputy Chief, who was assigned as the Shift Commander (Battalion 1) arrived approximately one minute behind Engine 10. The Battalion 1 vehicle was positioned on South Charles Street near the rear driveway entrance to establish a Command Post. The Acting Deputy Chief walked to the intersection to confer with the Lieutenant of Engine 10 while his aide set up the command post. He agreed with the Lieutenant's assessment of the situation and approved the attack plan, which was an interior attack from east to west to keep the exterior fire from extending into the building.

The Acting Deputy Chief then announced that he would be assuming command on the Charles Street side of the fire. His initial report was "Fire in a two-story, 50 x 80 foot building. Engine 5 will be operating on the exterior only from the rear. Engine 10 will be attacking through the interior from the opposite side. Battalion 1 will be Charles Command." The Lieutenant of Engine 10 was designated as Division B and assigned to supervise the companies on the east side of the fire.

The Incident Commander then assigned Engine 5 to the west side with specific instructions to prevent exterior spread of the fire, but not be make a direct attack from the west because the interior crews would be working from east to west. Engine 5 placed their manifold at the west driveway and stretched a 3-inch line to the hydrant at the southwest corner of Seventh Avenue South and South Charles Street.

Engine 13 was assigned to work with Engine 10 on the interior attack from the east side. The Lieutenant of Engine 10 assigned two of his crew members to work under the Lieutenant from Engine 13, while he continued to direct operations from the outside. Two 1-3/4 inch attack lines were advanced through the roll-up door into the south half of the building, one by Engine 13 and one by the crew members from Engine 10.

Engine 2 was assigned to work with Engine 5 on the west side. The Captain in charge of Engine 2 was designated as Division C and assigned to supervise the companies on the west side of the building.

Battalion Chief 5 arrived and was assigned to relieve the Lieutenant of Engine 10 as Division B. The Lieutenant and another crew member from Engine 10 then took a third line into the building through the office door.

Engine 36 reported to the Command Post and was assigned to report to Division B. They were directed to take another attack line through a door into the north half of the building to check for extension.

Ladder 3 approached from the east on South Dearborn Street and viewed the fire from the north. The Captain decided to place the ladder truck in the parking lot to the north of the fire building and to ladder the north wing. The crew then went to the roof with ventilation equipment. The deployment of first alarm companies is illustrated in Figure 2.

The Incident Commander requested two additional engine companies and one more ladder company to respond at 1910 hours and directed the incoming units to establish a Base at South Charles and Maynard. Four minutes later he upgraded the request to a full second alarm.

INTERIOR OPERATIONS

The first two attack lines were advanced into the building by the crews of Engines 10 and 13. They found the interior to be hot and heavily charged with smoke, but did not encounter any major interior fire involvement. A few spot fires were found and immediately knocked down with bursts of water from the hoselines. Most of the spot fires were reported to be near the floor level.

The crew advancing the third line encountered similar conditions as they worked their way through the interior. One spot fire was found near the floor level in the office area and others were found toward the rear. Some fire was also observed and extinguished on the underside of the roof. The crews were working in zero visibility conditions and had to advance slowly, moving around equipment and stored materials inside the building. They reported that the interior temperatures were hot enough to keep them crouched down, but not hot enough to cause unusual concern.

The three attack teams used up their air supplies and rotated out of change air cylinders. The attack lines were backed-out to the doorways where they were picked up by fresh teams and teams that had already exchanged their cylinders. Each team entering or leaving the building reported to Battalion Chief 5, who was maintaining the accountability system for Division B.

Engine 36 worked their way through the north wing, from east to west and encountered only one small spot fire. (This was near a small breach in the fire wall where the fire had communicated through from the basement.) They reported to Division B that there was no evidence of additional fire involvement in the north wing at that time.

WEST SIDE OPERATIONS

Engine 5 forced entry to the bakery area through the rear door and found no evidence of fire in this area. There was no smoke or heat and everything appeared to be normal in this area. The crew then raised a ladder to the roof of the west wing and took a 1-3/4 inch line to the top of the ladder. They did not go onto the roof because of a downed power line that was in their path. The hoseline was operated from the ladder to knock down the fire on the exterior of the west wall.

When the Captain of Engine 2 was assigned as Division C, he directed his crew to advance a back-up line for Engine 5 and then to take another handline around the west wing and into the north wing. They entered through the doors to the lower level of the north wing and found the area charged with smoke, but no evidence of fire or elevated temperatures.

ROOFTOP OPERATIONS

The crew of Ladder 1 went to the roof and found fire lapping over and onto the roof from the west side. They initially selected a point to open the roof toward the west side of the building, but the heat was so intense that they had to back away before the hole was cut. They decided to move back several feet to the east and to begin a north-south trench cut to reduce the risk of interior fire spread

Locations Of 1st Alarm Companies

Figure 2

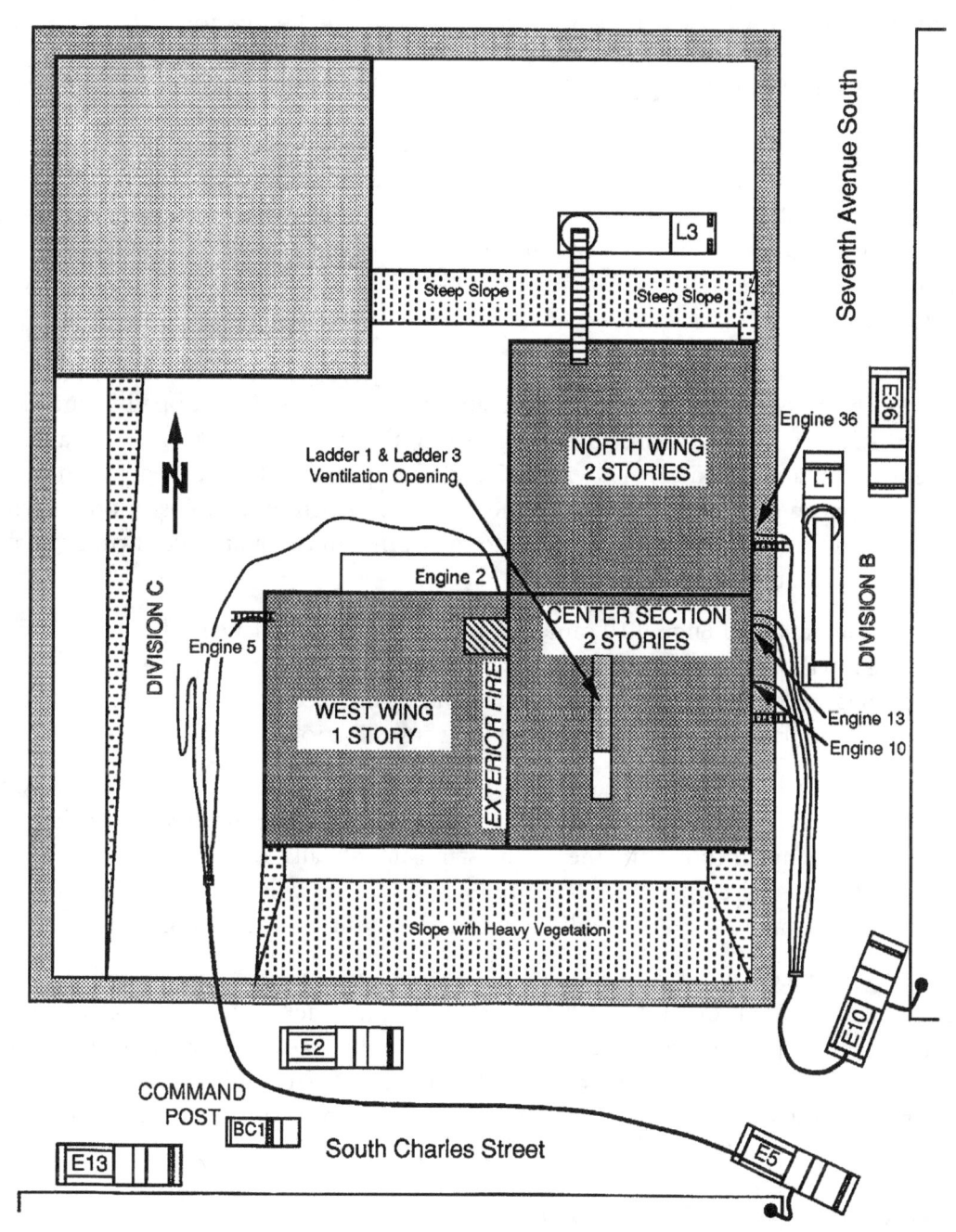

from west to east. They also asked Division B to send up a hoseline to protect them and to control any fire spread on the roof covering. The two crew members from Aid 5 (a BLS ambulance) were sent to the roof with an additional hoseline.

Ladder 1 was joined on the roof by the crew from Ladder 3 and both crews worked on the trench cut. The progress on this cut was slow because of the two layers of roof boards, one on top and one below the roof joists. They had to cut the roof covering with a chain saw and strip away the top layer, then reach down between the joists with the chainsaw to make a second cut through the lower layer of boards, then finally remove the lower boards to ventilate the interior.

CONTINUING INTERIOR OPERATIONS

The initial entry crews depleted their air supplies and had to come out to replace their air cylinders between 1920 and 1925 hours. Division B requested additional companies from the Incident Commander to rotate on the attack lines as the companies came out of the building. Ladder 7, which had responded on the special call, was directed to report to Division B and was assigned to take over one of the interior attack lines. Engine 36 and Ladder 3 were also assigned to take turns on the interior attack lines.

While en route to obtain a full air cylinder from his apparatus, the Lieutenant of Engine 13 observed the hoseline that was being operated by Engine 5. He stopped at the Command Post to express his concern that this line was pushing the fire toward the interior attack crews. The Incident Commander contacted Division C and reemphasized that the operation of exterior lines must not oppose the interior operation; he then left the Command Post momentarily to personally check on this operation. He returned to the Command Post after confirming that the exterior operation was not causing a problem.

When the Lieutenant of Engine 10 came out of the building to change cylinders, a few minutes later, he reported to the Incident Commander that this structure was the one they had been warned about as an arson target and to be sure that the investigators had been notified. (The on-duty investigators were already on the scene and had initiated their investigation.)

After changing air cylinders the crews of Engines 10 and 13 reported back to Division B and were sent back into the building to resume the interior attack. When they reentered they noted that the atmosphere was much cooler and they could stand-up to walk around. The roof had been vented by this time, which relieved the heat. The building was still heavily charged with smoke and they had difficulty navigating; however, they did not find any significant interior fire involvement.

Engine 13 advanced all the way to the west wall and into the lunchroom, where they found that the fire had burned through the wall from the exterior. They conducted overhaul in this area and attacked some fire that was visible through the hole in the wall. Ladder 7 advanced a line toward the middle of the floor area searching for fire, while the Lieutenant and one crew member from Engine 10 were working to their right with the third line. (The approximate positions of the interior attack lines are illustrated in Figure 3.)

BASEMENT FIRE DISCOVERED

The crew of Engine 2 had been assigned to look for additional areas where the fire could extend to the north and west. They located an opening in the north wall of the west wing that provided access to the interior loading dock. (The opening was an old doorway that had been covered over, but the covering was loose.) They advanced their line through this opening and used a step ladder that happened to be there to drop down to the loading dock.

When they reached the loading dock they found the sliding fire door leading into the storage room partially open. The interior of the storage room was fully involved in fire; however, no smoke or flames were coming out of the doorway. The large fire was drawing air in through this opening. (This location is shown in Figure 4.)

The crew advised their Captain (who was in charge of Division C at that time) of this discovery and discussed the possibility of launching an attack into the fire area from the west side. They were aware that the attack plan had been defined as east to west and that their assignment was to prevent further extension to the west. Following this plan they took positions to hold the fire at the loading dock, in anticipation of an interior attack coming toward them.

Battalion Chief 2 had arrived and reported to the Incident Commander at approximately 1925 hours. He was briefed on the attack plan, emphasizing that the objective for units on the west side was to prevent extension while the interior attack was made from east to west. He was then assigned to relieve the Captain from Engine 2 as Division C.

Battalion Chief 2 went to the west side of the fire building, was briefed on the situation by the Captain of Engine 2, then took charge of Division C. Engine 25 was assigned to report to Division C as a relief company.

At approximately 1930 hours it was recognized that the division designation were incorrect. The east side of the fire was redesignated as Division C and the west side was redesignated as Division B.

SAFETY OFFICER

The Deputy Chief of Training and Safety had responded from his residence on the working fire notification and assumed the role of Incident Safety Officer at 1922 hours. He initially conferred with Division B (Battalion Chief 5), verified that accountability procedures had been established, and evaluated the interior conditions from the east side. By this time the roof had been opened and there was no indication of a heated interior atmosphere. He noted that the interior was heavily charged with smoke and expressed concern to Division B that there could be a concealed fire that might suddenly break out or flash over. He emphasized the need to prepare a 2-1/2-inch back-up line and to assign a crew to stand by as a Rapid Intervention Team.

The Safety Officer then proceeded to the Command Post where he was briefed on the attack plan and reported his observations to the Incident Commander. He then continued around to the west side to evaluate conditions. He entered the west wing through the bakery door and noted the absence of smoke or any indication of fire in the lower level. He also noted that most of the exterior fire, which had been evident when he arrived, had been knocked down by Engine 5's hoseline.

He then encountered the Acting Assistant Fire Chief of Operations, who had arrived at the Command Post.

ASSISTANT CHIEF OF OPERATIONS

The Acting Assistant Fire Chief in charge of Operations had also responded from home on the initial report of a working structure fire in the International District. He approached from the north and parked his car northeast of the fire. From the corner he noted a strong thermal column rising from the rear portion of the structure. When he reached the building and looked in through the doors from the east side, he noted that the interior was heavily charged with smoke, but not hot. The inconsistency of these observations caused him to believe that there must be a significant fire burning somewhere in the structure, but that it must be in a different part of the building or in a concealed space. He asked the Division B Chief if the building had a basement and received a negative reply – this caused him to suspect a cockloft fire.

He then proceeded to the Command Post, where he discussed his concerns about a concealed space fire with the Incident Commander. The Incident Commander had not received a report from the roof and was not aware of the progress that was being made on vertical ventilation. The Acting Assistant Chief recommended calling for a third alarm, anticipating an extended operation, and told the Incident Commander that he wanted to make a full personal reconnaissance survey before assuming command of the incident. The third alarm was requested at 1932 hours.

The Acting Assistant Chief of Operations then encountered the Deputy Chief of Training and Safety, who reported that there was no evidence of fire in the lower level and that he had seen no major problems on the west side of the fire. The two of them returned to the east side and went to the roof to evaluate conditions, looking particularly for evidence of a cockloft fire.

On the rooftop they noted that a moderate amount of smoke was coming from the vents. This observation was inconclusive with respect to their suspicion that there could be a significant fire in the cockloft or in some other concealed space. They recognized that additional ventilation would be needed to either locate a concealed space fire or confirm that there was none.

The Acting Assistant Chief contacted the Incident Commander at 1936 hours to notify him of the need to assign a division supervisor to the roof. The Captain of Ladder 3 was assigned this responsibility pending the assignment of a command officer.

FLOOR COLLAPSE

At this time the three interior lines were being operated by Engines 10 and 13 and Ladder 7. The lieutenant and three firefighters from Engine 13 were operating in the northwest corner and the lunchroom. The lieutenant and one firefighter from Engine 10 were working in the north half of the ground floor, while the lieutenant and three firefighters from Ladder 7 were near the middle of the large space. All of the crews were in dense smoke, but the atmosphere was cool and there was no visible fire. The Lieutenant of Engine 10 and his partner briefly encountered the crew of Ladder 7, then disappeared back into the smoke.

Seconds later the building rumbled and flames erupted from the basement as the floor began to collapse. It appears that the "pony wall" failed, dropping the ends of the floor joists. Sections of the wood and concrete floor hinged down into the basement. The flames coming from the basement spread across the underside of the roof and the contents of the ground floor began to ignite in a rapid flashover sequence.

Two firefighters from Engine 13 were in the lunchroom and heard their lieutenant shout "…Let's get out of here!" as the floor began to drop. The two firefighters were able to go out through the hole in the wall and onto the roof of the west wing. The lieutenant and the other firefighter are believed to have fallen into the basement as one of the first sections dropped.

The Lieutenant of Engine 10 and his partner became separated as the floor collapsed. The firefighter was able to make his way back to the door and out, while the lieutenant dropped into the basement.

The crew of Ladder 7 at first believed that the ceiling must have collapsed as flames spread across the open area over their heads. The lieutenant opened the nozzle, attempting to cool the overhead, but the water stream immediately turned to steam. The four crew members began to follow their line back toward the door in single file as intense heat radiated down on them. They did not realize that the floor was collapsing until they encountered flames coming up through a large opening, almost directly in their path back to the doorway.

Two of the firefighters and their lieutenant managed to scramble to the door and outside, passing within feet of the opening. When they reached the exterior they realized that the third firefighter was no longer with them. The missing firefighter had been first in the line as they were trying to find their way out and he either fell into a hole or dropped into the basement as a floor section collapsed under him.

All of the seven firefighters who escaped were burned. The majority of the burns were to their necks and ears. The Lieutenant of Ladder 7 had additional burns to his wrists and one hand.

ACCOUNTABILITY

It was evident within second that something was going wrong; however, personnel outside the building were not immediately sure what was happening. Hot heavy smoke and some flames began to issue from the doors and out through the hole in the roof. On the east side of the building firefighters began to scramble out the door with their protective clothing smoking. It took almost a full minute before the Lieutenant of Ladder 7, the last to escape, came out of the building.

On the west side the two firefighters from Engine 13 suddenly appeared at the top of the ladder, having crossed the roof of the west wing, reporting that their lieutenant and another firefighter were missing.

The radio came alive with messages to abandon the building and the evacuation tones were sounded by the Incident Commander. The Communications Center reported that an emergency notification signal was being received from one portable radio, then from a second radio. As soon as he could get outside and account for his crew, the Lieutenant of Ladder 7 transmitted a message that one of his crew members was missing.

It took only a few minutes to account for all of the crew and crew members and to confirm the identities and last known locations of the four missing personnel. Crews outside the building were quickly organized to operate streams into the opening in hopes of protecting the firefighters who believed at this point to have fallen into the basement. Search and rescue plans were developed, based on their last known locations.

RESCUE ATTEMPTS

The Acting Assistant Chief of Operations assumed command of the incident and assigned the Acting Deputy Chief to manage the Operations Section. The 4th and 5th alarms were transmitted for additional resources and four additional medic units were requested.

Two Rescue Branches were established, one on the east side to conduct rescue operations on the upper level and one on the west side to make a similar effort in the lower level. The floor collapse was determined to involve only the sections immediately south of the interior fire wall, so there was a possibility that some of the missing firefighters could still be on the upper level. After evaluating structural conditions crews were reassigned to the roof to provide additional vertical ventilation.

Figure 3

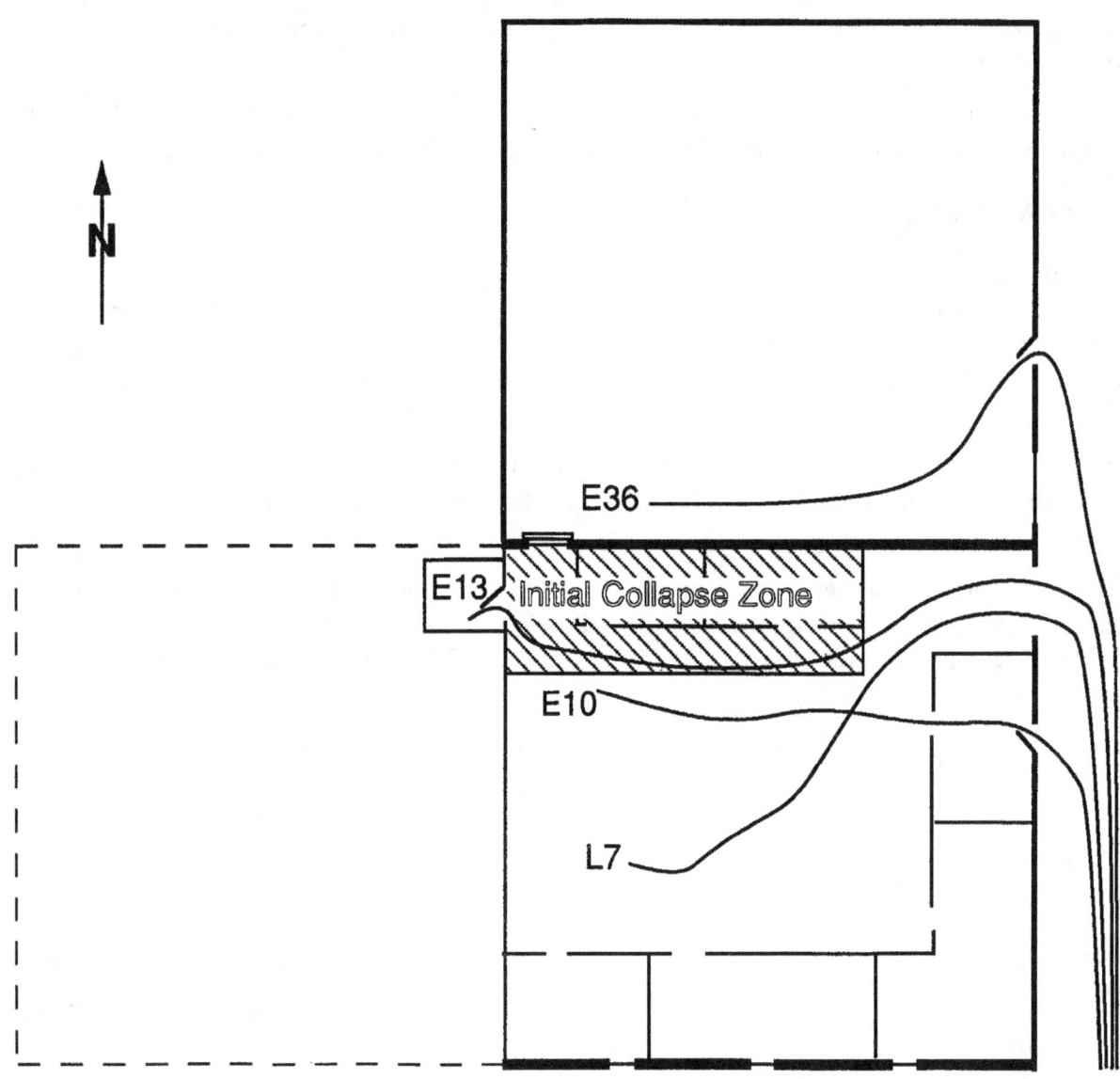

UPPER LEVEL FLOOR PLAN
Showing Attack Line Positions at Time of Collapse

N

E36

E13 Initial Collapse Zone

E10

L7

4 Attack Lines
All 1 3/4 inch
Supplied by
E10 Manifold

Figure 4

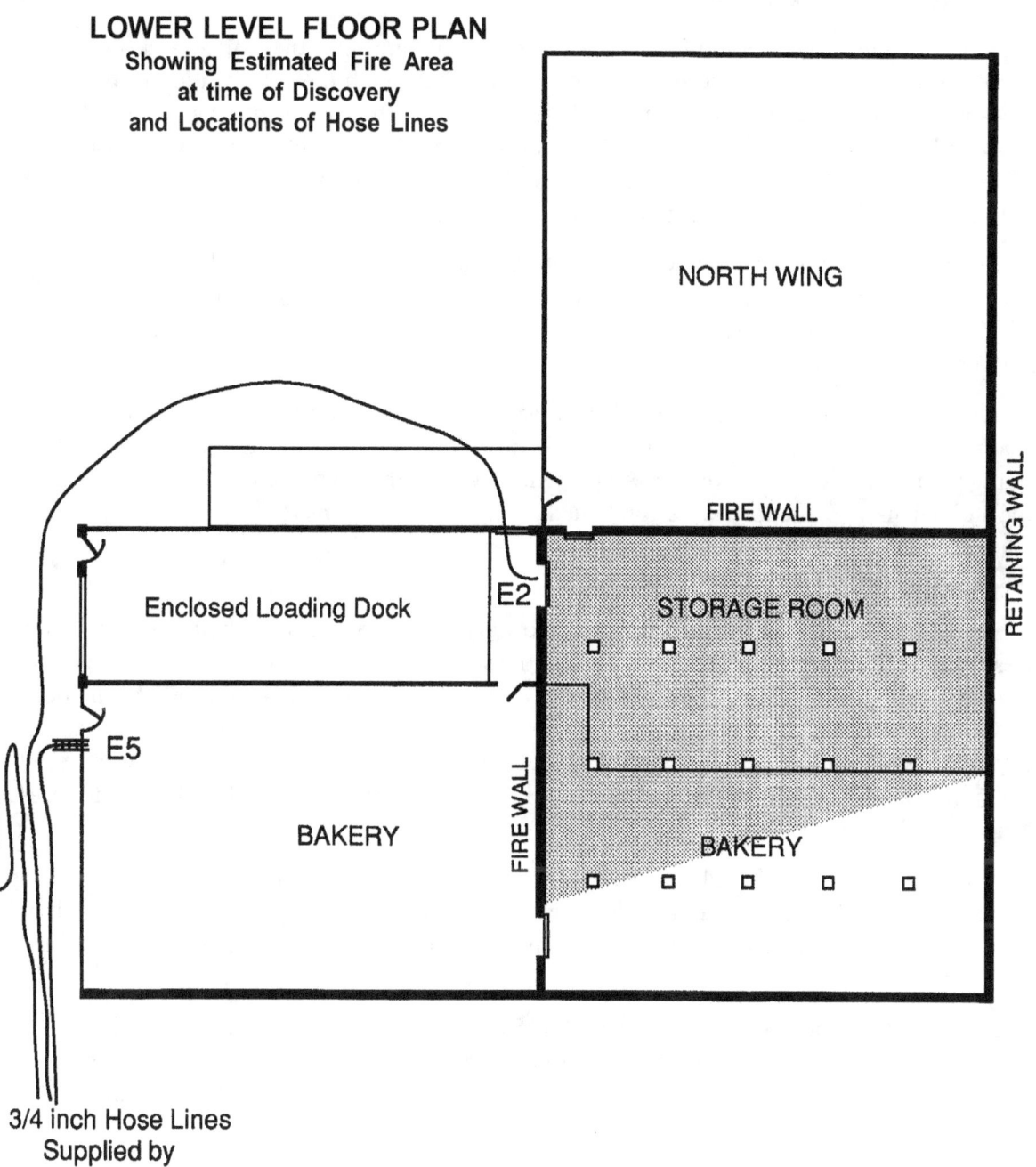

LOWER LEVEL FLOOR PLAN
Showing Estimated Fire Area
at time of Discovery
and Locations of Hose Lines

NORTH WING

RETAINING WALL

FIRE WALL

Enclosed Loading Dock

E2

STORAGE ROOM

E5

FIRE WALL

BAKERY

BAKERY

1 3/4 inch Hose Lines
Supplied by
E5 Manifold

The crews upstairs had very little success penetrating the structure, due to the intense heat; however, the crews in the basement made several entries including some deep penetrations into the rubble. With the openings in the floor above and in the roof the heat and smoke were venting and the rescue teams were able to work their way into the rubble.

Engine Companies 2 and 25 and Ladder 10 advanced hoselines into the storage area and attempted to move some of the debris, but they were unable to get through the combination of burning contents and materials that had fallen into the basement. They then went in through the bakery area, which had also become involved in the fire, and worked their way all the way to the east wall. At different times they believed that they could hear PASS units sounding or SCBA low pressure alarms, but they were unsuccessful in locating any of the missing personnel.

The rescue attempts involved a very significant personal risk to the rescuers; however, they were all well organized and very conscious of conditions. It was later determined that the rescuers came within a few feet of two of the victims, but they could not be located in the rubble and probably could not have been saved at that point.

The rescue efforts were discontinued when it was determined that the risk of additional structural collapse was imminent. At this point victims had been missing for over an hour and any hopes of finding them alive had been given up. All personnel were withdrawn from the structure for a second time and defensive operations were conducted, using master streams to control the fire.

BODY RECOVERY

Body recovery efforts were initiated the next morning, in coordination with an intensive fire cause investigation. The bodies were believed to be in the storage room, which was also the suspected area of origin of the fire. This situation necessitated a very slow and deliberate removal of the debris to recover and document evidence, while also searching for the bodies. The structure had to be partially demolished and parts had to be braced before entry could be made. Two bodies were recovered during the first day, one on the second day, and the last body was not located until the third evening, 72 hours after the fire was reported.

All four firefighters are believed to have died from asphyxiation after running out of air or losing the integrity of their SCBAs when they fell. Two were incapacitated and died where they fell, while two had managed to move a considerable distance from the points where they are believed to have fallen into the basement.

All four of the firefighters were equipped with PASS devices and all four also had portable radios. The emergency buttons on two of the radios had been activated and two of the PASS devices had been activated. The other two PASS devices were found with the switches in the OFF position.

INCIDENT ANALYSIS

The deaths of four firefighters resulted from a combination of factors and circumstances. The direct cause of the fire and the resulting deaths was an act of arson. Notwithstanding this criminal act and the associated legal responsibility for the four deaths, this section of the report examines several contributing factors that relate to the fire suppression operations and the cause of the structural collapse.

STRUCTURAL COLLAPSE

The sudden floor collapse can be attributed to an unusual construction detail which would have been extremely difficult to anticipate or predict. A visual examination of the major structural elements of the building would have suggested that they had fire endurance characteristics similar to a heavy timber building. The "pony wall" was an undetected weak link in the structural system.

The presence of the "pony wall" could not have been identified during the fire, because it was inside the room that was burning, and could only have been anticipated through some prior knowledge of its existence. This would have required accurate construction plans for the structure, which were not available, or a professional inspection by an individual with expertise in structural analysis. It was inconsistent with the structural details that could be observed in other parts of the structure.

PRE-FIRE PLAN

The Seattle Fire Department did not have a pre-fire plan for this building. The preparation of pre-fire plans is a company level responsibility and company officers determine the need to prepare plans and conduct familiarization tours of individual properties in their areas. This building had not been identified as requiring a plan, because it had not been recognized as an unusually complex or dangerous occupancy.

A pre-fire plan might have assisted the Incident Commander and other officers in understanding the complicated arrangement of the fire building. Although one of the objectives of a pre-fire plan is to make officers aware of hidden dangers, it is doubtful whether the presence and significance of the "pony wall" would have been noted by a company assigned to prepare a plan of the Mary Pang building. It would have been very difficult to observe, because it was located in a windowless area, high above the floor, against a brick wall, in a room that was used only for storage. The individuals assigned to conduct the pre-fire survey would have to be exceptionally diligent to locate the "pony wall" and recognize its significance as a weak link in the structural integrity of the building.

Construction plans are sometimes available to assist firefighters in preparing pre-fire plans. However, there were no accurate "as built" plans for the structure available at the site or on file with the City of Seattle. The building is known to have been modified several times over its estimated life of 85 years, and there were no accurate records of the many changes that had been made over the years. The plans that were on file with the city did not show the "pony wall" or indicate when it might have been installed. Similar conditions may exist in similar structures in the same area.

INCIDENT COMMAND SYSTEM

The Seattle Fire Department uses the Incident Command System (ICS) to manage its emergency incidents. The ICS system was applied in a standard and consistent manner from the outset of operations and the incident can be described as well organized and well managed, "by the book." The strategic plan and specific assignments were communicated very effectively and supervisory responsibilities were assigned in a standard manner. There was a delay in establishing a separate division for rooftop operations; however, this was recognized and corrected and had no impact on the outcome. The rescue and recovery phases of the incident were exceptionally well managed.

A problem was noted in the flow of information and progress reports back to the Incident Commander. This relates to the number of functions that are assigned to Division Supervisors and the impact these responsibilities can have on their ability to actually supervise and monitor tactical operations. This is discussed in more detail in the following section of this report.

There was some confusion in the analysis of this incident with regard to the convention for naming the Divisions. The Seattle Fire Department uses the A-B-C-D system for designating the sides of a building; however, the designations were not assigned in the standard manner and were changed during the course of the incident. There is no indication that the sector designations were a significant factor in the incident, because most of the companies reported to and received their instructions at the Command Post; only one company reported momentary confusion over where they were supposed to report. This could be a more significant problem in other situations and it points out the value of using a uniform system for every incident.

RISK ASSESSMENT

The importance of conducting fire suppression operations with a "risk management approach" is emphasized in NFPA Standard 1500, the Standard for a Fire Department Occupational Safety and Health Program. The application of the risk management concept to this incident is particularly significant.

The Incident Commander is responsible for defining the overall strategy that will be used at an incident and for the development and execution of a strategic plan. A strategic plan must be based on a combination of factors, including a size-up of the structure and the fire conditions and a risk assessment of the situation.

The Incident Commander seldom has complete knowledge of all the potential risk factors during the early stages of an incident. Operations must often be initiated based on the best information that is available and then adjusted, if necessary, as additional information is obtained.

The Incident Commander's initial risk assessment must be based on the information that is available at that time. Before initiating an offensive operation, the Incident Commander must make a positive determination that an interior attack is reasonably safe. If the risk assessment does not support this positive determination, the Incident Commander must select a defensive strategy and limit operations to the exterior of the structure.

At this incident the situation gave the appearance of a heavy timber structure, occupied by two operating businesses, that was exposed to an exterior fire. The building had been occupied by tenants when the fire was discovered and there could still have been occupants inside. The initial risk assessment supported an interior offensive attack to keep the fire out of the building and an east to west attack plan was effectively communicated to the operating crews. This plan was identified by the first arriving officer and approved by the Acting Deputy Chief who assumed command from him. Additional companies were assigned to the west side to prevent extension of the fire and specifically directed to avoid a conflict with the interior attack.

Risk assessment is a continuous process that is directly dependent upon information management. As the situation progresses and additional information is obtained, the Incident Commander must regularly and routinely reprocess the risk assessment. If the risk assessment changes the Incident Commander may have to change the strategic plan. When the risk assessment changes from positive to negative, offensive strategy must shift to defensive and attack crews must be withdrawn from unsafe positions. A shift from defensive to offensive strategy when the risk assessment improves is discretionary.

As this incident progressed, the attack plan appeared to be working. The visible exterior fire was controlled and there were no reports of significant interior involvement. The fire in the basement had been located, but it was not reported back to the Incident Commander. This was critical information that could have changed the risk analysis and caused him to evacuate the interior crews before the floor collapsed. However, the significance of the information was not recognized and it was not communicated to the Incident Commander.

Information Management

The ability of an Incident Commander to make an appropriate risk assessment is necessarily based on the information that is available at any point in time and the Incident Commander's ability to process that information. The Incident Commander must use the available information in an expedient manner to direct the early stages of an operation and then continually update the information.

The initial information incorporates observations of the structure, visible smoke and fire conditions and other situational factors, which can be processed along with the Incident Commander's general knowledge and experience to develop an initial strategic plan. This process can be greatly enhanced by the availability of a pre-fire plan for the occupancy and by information that has been obtained through prior familiarization tours and previous incidents.

Pre-Fire Plan

In this incident a pre-fire plan might have made the Incident Commander and other officers more aware of the arrangement of the structure. This could have resulted in an earlier recognition that the fire was in the basement and a decision to evacuate the crews that were operating above it. A different risk assessment would have been required to consider an interior attack into the basement. As noted previously, it is questionable whether or not the "flaw" in the structure would have been recognized from a pre-fire plan.

Observations

The Incident Commander's specific knowledge of the situation should improve as operations progress, based on personal observations, progress reports, and input from other individuals and sources. As the information improves and the incident evolves, the Incident Commander's ability to assess all of the potential risk factors should improve.

It is often very difficult for the Incident Commander to personally observe changing conditions at complex incidents. The observations of others are important input to the process. All officers should be continually observing conditions and making personal risk assessments. They should immediately inform their superiors or the Incident Commander if they become aware of any information that could impact on the overall strategy for the incident. This process must be efficient and messages must be prioritized to avoid "information overload" and to ensure that significant information is transmitted without delay.

Progress Reports

One of the primary responsibilities of company officers and division or sector officers is to keeping the Incident Commander informed. They should make regular progress reports that include any observations that could be pertinent to the Incident Commander's risk assessment and strategic planning. The Incident Commander often has to depend on progress reports and information from other observers to evaluate the effectiveness of the attack plan.

At this incident the assigned Division Supervisors did not transmit any progress reports, positive or negative, to the Incident Commander. Division Supervisors have several important responsibilities to addition to regularly reporting conditions and progress back to the Incident Commander. They should be constantly aware of the conditions that are being encountered and the progress that is being made by their assigned companies.

To evaluate interior progress, Division Supervisors must either go inside, where the companies are operating, or depend on the company officers to keep them informed with accurate reports. In this case the company officers who were leading the interior attack teams did not provide any progress reports or information on interior conditions to their Division Supervisor, who was located outside the building. He could see that the interior was heavily charged with smoke, but he did not know that the companies were encountering very little fire inside. He did not have any information to provide progress reports to the Incident Commander and was not asked for a report.

The Division Supervisors on the west side saw the large fire in the basement a few minutes before the collapse occurred; however, the significance of this observation was not recognized and it was not reported. The personnel on the west side assumed that the Incident Commander knew about the fire in the basement and did not realize that the interior attack crews were working above it.

A progress report from either of the two Divisions could have caused the other Division Supervisor to recognize the inconsistency and alert the Incident Commander to the problem.

The lack of progress reports appears to be related to the lack of aides or assistants to support the Division Supervisors.

When Division Supervisors cannot directly monitor the work that is being performed by their assigned companies and the fire conditions in their assigned areas, there is a risk that several important functions, including supervision, coordination, safety surveillance, and progress reporting will be compromised. This can occur when the Division Supervisors are overloaded with responsibilities.

Seattle uses a "passport" accountability system, which requires an Accountability Control Point to be established outside the operating area, near each entry and exit point. The Division Supervisors are responsible for this function, as well as for monitoring two separate radios, one on the tactical channel and one on the emergency channel. These important responsibilities tend to keep the Division Supervisors outside, at secondary command posts. They must rely on company officers to perform the interior supervision and coordination functions and to keep the Division Supervisors informed about what is happening inside.

Some of these responsibilities could be delegated. However, the Battalion Chiefs, who are normally assigned as Division Supervisors, do not have aides. Also, because of the rule that a minimum of two personnel are required to make an interior entry, a Division Supervisor would need a partner to be able to go inside. To fulfill all of the requirements and allow the Division Supervisor to enter the structure, two assistants would have to be assigned to each Battalion Chief. This would require a company or other available personnel to be assigned to support each Division Supervisor.

In the absence of progress reports from the Division Supervisors, the Incident Commander had to rely on other indicators to evaluate the effectiveness of the attack plan. The exterior fire was knocked down, which appeared to indicate that the fire was being successfully controlled. He received reports from other officers indicating that the interior was smoke charged, but not hot, and that there was no sign of smoke or fire in the basement. All of the reports suggested that the strategic plan was working. The observation that there could be a working fire in a concealed space only reached the Incident Commander a few moments before the floor collapsed.

Reconnaissance

Visual reconnaissance by the Incident Commander or by other individuals who are assigned to report back to the Incident Commander is an additional source of valuable information. At this incident it was difficult to visually size-up the situation from a single vantage point. The unusual configuration of the building and the surrounding grade levels made the structure difficult to interpret without a pre-fire plan or a 360 degree size-up. The differences in grade levels made it difficult to lap all the way around the structure to make a better size-up. It would have been impossible to evaluate the smoke filled interior without going inside.

The visible smoke and fire conditions were difficult to relate to the actual location and magnitude of the fire inside the structure. All except one of the companies and command officers responding on the first alarm approached the fire from the same direction, eastbound on South Charles Street from Maynard. All of the officers and many of the crew members all saw the large volume of flames against the west wall of the structure, which they all interpreted as an exterior fire threatening to extend into the building.

From the Command Post the view was obstructed by trees and vegetation which made it difficult to visually size-up the structure or the fire conditions. The area where the basement fire was venting to the exterior was below the street level and out of sight from the Command Post.

The interior attack crews never encountered any significant interior fire involvement on the upper level; they found only a few spot fires near the floor which they quickly controlled. The initial conditions were heavy smoke and moderate heat, which was consistent with their expectations. When rooftop ventilation was accomplished and the heated gases were released, the interior atmosphere cooled, which is usually an indicator of good progress. They did not realize there was a basement below them – the concrete floor reinforced this interpretation of the structure and prevented smoke and flames from penetrating through from below. They believed that their efforts were successfully keeping the fire out of the building.

The only company that approached from the north side, Ladder 3, went directly to the roof. Their observations suggested that the visible fire was coming from a separate structure located to the west and threatening to spread into the main building.

The companies assigned to the west side had the best vantage point to see the configuration of the building. Engine 5 operated their hoseline from a ladder over the roof of the west wing. They could not advance the line out onto the roof because of the electrical power line that had dropped in their path; this kept them from closely examining the west wall of the two story section, which would have allowed them to see that the visible fire was actually venting out of the basement.

When Engine 5's crew made an entry into the bakery, the atmosphere was cool and clear of smoke and provided no indication of an interior fire. All of the fire was on the other side of the brick wall and the heat and smoke were venting out through the higher opening.

The second company assigned to the west side, Engine 2, eventually located the door into the fire area and determined that a large area of the basement was heavily involved in fire. Because of the assignment they had been given, they expected the attack crews to be pushing the fire toward them, so they prepared to defend their position. They discussed the possibility of attacking the fire and determined that it would violate the strategy that had been announced for the fire. They did not recognize that no one else was aware of the large interior fire or that the interior crews were directly over it.

In retrospect it can be determined that very different interpretations of the structure and the fire were being made from different vantage points. The individuals making these observations each believed that their interpretations of the structure were accurate, but did not recognize the significance of their information to the Incident Commander.

This emphasizes the value of a complete 360 degree size-up of the fire scene, as early as possible, by the Incident Commander or by an individual who can report in person to the Incident Commander with a "full picture" of the scene. In the absence of personal observations or pre-fire plans, the Incident Commander must rely on others to provide information from their different vantage points, then must assemble and interpret their observations.

Additional observations were made by several other individuals who did not recognize their significance at the time:

- One of the rookie firefighters on an interior attack team kneeled on the concrete floor momentarily and noted that it was very hot. He made a mental note to ask his Lieutenant why the floor was so hot when they returned to the station.

- The firefighter operating the exterior line on the west side noted a bright red band of fire, after most of the exterior fire had been extinguished. This band was approximately 4 feet high, beginning at the point where the roof of the west wing met the vertical wall, and extended across the full width of the center section of the building. The firefighter was seeing the interior fire in the basement, below the upper floor level, from a distance of 60 feet. He pointed out the unusual observation to his company officer and to the Battalion Chief who was just arriving in that area to become the Division Supervisor; however, no one noted its significance at the time. This was approximately three to five minutes before the collapse occurred.

- The band of fire was captured in a photograph that was taken by an investigator; however, it was not recognized until the film was developed.

- Some of the interior attack personnel saw one or more small fires at the floor level in different areas of the warehouse; however, they did not report this observation until later. Some of the wood frame partitions had been installed before the concrete floor topping was added and were nailed directly to the wood floor boards. It appears that the fire was extending into these partitions through gaps between the floor boards.

The Acting Assistant Chief of Operations made the observation that the interior conditions on the upper floor were inconsistent with the thermal column that could be seen from a block away. This caused him to suspect a concealed space fire and to conduct a personal reconnaissance before assuming command of the incident. He also informed the Incident Commander of this suspicion. The comparison of indicators from different vantage points often reveals significant information about a fire. It is important to look at the "big picture" as well as individual narrow views.

Safety Officer

Incident Safety Officers play an important role in managing safety and supporting the Incident Commander in the risk assessment process. The Seattle Fire Department routinely assigns designated command officers as part of the command organization for working incidents. The Deputy Chief of Training and Safety and the Battalion Chief who is assigned as the Fire Department Safety Officer both responded to this incident. The Deputy Chief arrived first and assumed the position of Incident Safety Officer. The Battalion Chief had just arrived and was in the process of assuming a second Incident Safety Officer position when the collapse occurred.

The initial Safety Officer began an initial tour of the incident scene to identify problems and hazards and report back to the Incident Commander. He first entered from the east and noted that the interior smoke was heavy, but the atmosphere was not hot. He was concerned that a concealed fire might suddenly break out and noted the need for a back-up line and a Rapid Intervention Team.

He then went to the west side and looked into the lower level from the bakery door. He noted that there was no evidence of smoke or heat in this area and returned to the Command Post to provide an initial safety report based on these observations. If he had continued around the corner of the building he would have encountered the crews that had discovered the fire in the basement.

At the Command Post he met the Acting Assistant Chief of Operations, who had made the same observation of the interior conditions on the upper level. This caused them both to suspect a concealed space fire. Believing that the basement was clear, they went to the roof to check for evidence of a cockloft fire. They were still on the roof when the floor collapse occurred. If they had been able to complete a 360 degree survey of the fire scene, they would have reached the location where the fire in the basement was visible and recognized that the attack crews were operating above it.

ACCOUNTABILITY

The Seattle Fire Department has placed a major emphasis on developing and implementing its accountability system, which is an important component of its operational safety procedures. Seattle uses a "passport" system for personnel accountability, which incorporates the following features:

- Each individual has at least two personal identification tags with hook and pile backing. When the member is on-duty, one of the personal tags is placed on the company "passport" – a small board that accommodates a tag for each crew member.

- Each company officer carries the passport for the company when reporting in at the scene of an incident and an identical backup passport is left on the apparatus. The passport is left with a designated individual near the entry point when the company enters a hazardous area and is retrieved when the company exits.

- The designated individual depends on the chain of command for the incident. At small scale incidents the passports are placed on the Incident Commander's control board. When divisions are established, the passports for all companies working in the Division are placed on the Division Supervisor's control board – the Division Supervisors' passports are placed on the Incident Commander's control board.

- One or more individuals may be assigned to manage the accountability function for the entire incident or at each entry/exit location. The designated individuals keep track of the times that companies enter and exit. All company members must enter and leave together.

- Companies with four or more entry personnel may be split into entry teams – each team must have a minimum of two personnel. Each entry team has its own passport and is identified with a letter (example: Engine 10A and Engine 10B).

- Each individual's helmet is marked with their company or unit number. The markers are attached to the helmets with Velcro and may be changed immediately if the individual is reassigned, even if it is only for a brief period.

- A current duty roster is maintained on the Computer Aided Dispatch (CAD) System for each unit. Company officers update the rosters at the beginning of each shift and whenever the assignments change during a shift. When a company or vehicle is dispatched to an incident, a roster of all the personnel assigned to that unit is printed out by the CAD system for the command officer. The same information is available at the Communications Center and can be transmitted to the mobile command post vehicle.

- Each individual on-duty is assigned a portable radio which has an emergency button and a unique four-digit identification number. When an emergency notification button is activated, the radio switches to the emergency channel and the ID number is displayed at the Communications Center. The name and assignment of the individual are referenced from the daily roster.

- Each individual has an SCBA assigned for the shift and each SCBA has a PASS device attached to it.

The accountability system worked well at this incident, keeping track of companies and individuals throughout the incident. The assignment of a portable radio to every on-duty firefighter is a very significant advance in fireground safety and accountability. The radios were acquired as part of a countywide 800 Mhz trunked public safety radio system and had been in service for only a short time when this fire occurred. Two of the trapped firefighters managed to activate the emergency buttons on their radios and the signals were received and decoded at the Communications Center.

Two problems were noted with the new radio system. One of the emergency identifiers was decoded as coming from an individual who was not at the fire, apparently because the reference database had not been updated or a number had been entered incorrectly that day. This error was identified almost immediately when the roster was checked and the individual called in from a fire station where he was monitoring the radio traffic from the incident. The value of this technology was proven, in spite of the fact that it did not correctly identify both individuals.

The second problem is that the design of the new system requires two radio channels to be monitored by Command Officers, because emergency signals are transmitted over a special frequency. This is a system configuration problem that is being reevaluated as further development occurs with the radio system.

SAFETY EQUIPMENT

All of the personnel involved in this incident were wearing full protective clothing that complied with current standards and using self-contained breathing apparatus. There were no indications that the protective clothing or equipment failed to perform as designed. They were not wearing protective hoods, which could have provided somewhat better protection to their necks and ears; however, there is no indication that this would have changed the outcome of this situation in any manner. The Seattle Fire Department subsequently decided to purchase and issue hoods to all firefighters.

It appears that two of the firefighters who died had not turned on their PASS devices. This observation has been made in other investigations and it has been determined that many firefighters are not using these devices because of the nuisance value created by frequent false alarms. This issue is a current concern in several fire departments and fire service organizations.

LESSONS LEARNED AND REINFORCED

Hidden Structural Flaws

Buildings that appear to be structurally sound may contain hidden flaws that are very difficult to identify or anticipate. This incident has made the Seattle Fire Department aware of a particular construction detail that may be present in similar structures in the area. All fire departments should attempt to identify dangerous construction features that are likely to be encountered in their areas and then survey properties where they are likely to be present. This information should be documented in pre-fire plans and reinforced through training programs.

Risk Management

Fire departments must constantly balance the risks of their work environment with the mission they are called upon to perform. This incident clearly demonstrates the complexity of the risks that must be considered and the reasons that operational safety has become a priority for fire departments. The Seattle Fire Department is a recognized leader in the development of operational safety and accountability procedures and this incident was well managed with a very high regard for safety. In spite of these efforts, four firefighters died in a fire that was deliberately set in a building with an undetected structural weakness.

Pre-Fire Planning

The development of pre-fire plans for significant occupancies, particularly those that are complicated or may involve unusual risks, should be an important program for all fire departments. Critical information must be documented and compiled in a system that makes it immediately available when an incident occurs.

Incident Command

The importance of effective communications and information management in the Incident Command System are demonstrated in this incident. The strategic plan and tactical objectives were very effectively communicated by the Incident Commander; however, the flow of information and progress reports back to the Incident Commander was inadequate. Important pieces of information were known to several individuals at the scene; however, they did not recognize the significance of their information and it was not reported back to the Incident Commander.

Command Support Functions

As incident command, accountability, and risk management have become more complex and structured, the role of Command Officers has changed in many fire departments. The added responsibilities, which are important, may restrict their ability to actually supervise, direct, and coordinate tactical operations, particularly when they are not provided with aides or staff support personnel. This may occur when the Command Officers have to direct their efforts to personally perform tasks, such as maintaining accountability systems and monitoring multiple radio channels. It is important to ensure that the basic responsibilities are not abandoned in favor of functions that could be effectively performed by support staff.

Experience

The simple value of experience in observing and interpreting fire conditions can be seen in this incident. Two senior officers had recognized inconsistencies between smoke and heat conditions inside the building and those observed from a distance and were in the process of making a full reconnaissance at the time the floor collapse occurred. It is likely that they would have discovered the basement fire and recognized the extreme danger within a very short time – unfortunately, the fire did not provide that time. The value of good experience should never be discounted.

Accountability

The Seattle Fire Department's accountability procedures proved to be effective in rapidly determining who was missing and where they were last seen. This allowed rescue attempts to be quickly initiated and well organized.

Risk Management During Rescue Efforts

Firefighters are inclined to extend valiant efforts at tremendous personal risk in attempts to rescue other firefighters. In these situations it can be very difficult to balance standard approaches to safety with the willingness of rescuers to accept unusual levels of personal risk. It is extremely difficult to make the decision to discontinue search and rescue operations, even when there is no realistic hope for success.

Extended Operations

When operations are extended over a long period of time, it can be very challenging to maintain safety and accountability procedures, particularly when multiple agencies are involved and firefighter fatalities have occurred. These incidents require a high level of planning, coordination, and discipline. It is important to rotate crews and provide adequate rest, rehabilitation, and support for all personnel, including staff and command officers.

Multiple Investigations

Situations that require simultaneous fire cause and line-of-duty death investigations are complicated and require a high level of coordination among the responsible individuals. The complexity increases rapidly when criminal acts are involved and additional organizations and agencies become involved in the investigation process. Good preparation and planning, along with positive established relationships, are essential in these situations.

APPENDIX A

Structural Details

STRUCTURAL DETAILS

This section of the report provides additional details relating to the structure and the failure analysis. The figures on the following pages illustrate the structure described.

All of the major structural components of the center section of the fire building were either brick walls or heavy timber members. The heavy timber members have inherent fire resistance that can be attributed to their mass. The "pony wall" that supported the ends of the floor joists at the north wall was the only exception and can be identified as the "weak link" in the fire resistance of the structural system. The sudden failure of this "pony wall" released the ends of the floor joists and caused sections of the upper floor to drop into the basement without warning.

The roof of the original structure was supported by 12 by 12 inch wood posts and by the exterior brick walls. When the second story was added, the new floor was constructed with a 2-inch thick wood deck, supported on 3 by 14 inch wood joists, resting on 8 by 14 inch wood beams. The floor beams, which spanned from east to west, were supported on cast iron caps placed on top of the original 12 by 12 inch wood posts. The outer ends of the beams rested on pilasters at the east wall and on top of wood posts that were attached to the west wall. To support the new roof, 12 by 12 inch wood posts were added directly above the original posts, standing on top of the cast iron column caps.

The integrity of the entire structural system depended on gravity. There were no mechanical fasteners connecting the column caps to the floor beams or to the upper posts. The weight of the structure held all of the components together. There is no evidence, however, that any of the gravity connections failed and contributed to the sudden structural failure. Several of these connections failed later in the fire as the integrity of the entire structure deteriorated.

At the north wall the original roof had been supported by a narrow ledge that was incorporated in the brick wall. Because the new floor was considerably higher than the old roof, the "pony wall" was installed on top of the ledge to support the ends of the new floor joists. The "pony wall" was fabricated from 2 by 4 inch lumber which has much less fire resistance than any of the other structural members. At the south wall the floor joists were supported by pockets in the brick exterior wall, which was raised to the new roof level.

Above grade structure

Above grade the new structure was built with one brick wall (on the south side) and three wood frame walls. The three wood frame walls were constructed with wood posts, sandwiched between one inch thick "shiplap" boards, a common type of construction for the Seattle area at that time. These wood frame walls were supported by the new floor deck.

The new floor was approximately four feet higher than the roof of the single story west wing. This left a gap between the top of the original brick wall and the underside of the new floor. This gap was filled by an additional section of wood frame exterior wall. At a later date the entire west wall was covered by metal siding which obscured the construction details.

The new roof deck was tar and gravel on wood boards, nailed to wood joists, with a second layer of boards nailed to the underside of the joists. The double layer roof was very difficult to ventilate because the two layers had to be cut consecutively. The upper deck had to be cut with chainsaws and removed, then a second pass had to be made with the saws, reaching down into the opening to cut the inner deck.

Basement

With the change in grade elevations, the lower level of the center section became a windowless basement. The floor of this basement was almost twenty feet below the street level and there was a crawl space below the floor. All four walls of this basement were brick, except for the topmost section of the west wall, which was the frame "filler" section. The fire in the basement burned through this wall section and extended up the exterior of the wall above, before the first units arrived at the scene. This appeared to be an exterior fire to the units approaching from the west.

At some time a concrete floor topping, approximately one inch thick, was added on top of the upper level floor decking, which helped to confine the fire to the basement. The stairway that had connected the upper and lower levels was also permanently covered over and topped with concrete, eliminating any interior access between the floor levels. The only remaining access to the basement was through the west wing or through a single narrow door into the lower level of the north wing.

The basement of the center section was divided into two floor areas. A sliding fire door provided access from the loading dock to the storage room. The door to the north wing door provided a second access into the storage room, but this door was difficult to access and seldom used. The bakery section was accessible through a door from the portion of the west wing that was occupied by the bakery.

West Wing

The west wing remained as a single story structure and is considered to be the "rear" portion of the building. With the change in grade levels the west wing was actually below the grade level of the adjacent streets. A driveway and parking lot on the west side sloped down to a rear vehicle entrance that provided access to an interior loading dock. The floor level in the west wing is several feet lower than the driveway.

The west wing was also divided into two areas. The south portion was occupied by the bakery. The north portion was an interior parking area and loading dock. A ramp inside the building sloped down to the dock level; the roof over the ramp had to be raised to accommodate vehicles. Several vehicles were parked in this area.

North Wing

The north wing was occupied by the Mary Pang Chinese Food Company. An upper floor was also added to the north wing; however, it appears to have been added at a later date than the upper level of the center section. When it was added, the brick wall between the center section and the north

Appendix A (continued)

wing was extended upward to create a fire separation, leaving the wood frame wall of the center section in-place. The north wing was extensively remodeled in the 1970's.

The lower floor in the north wing was several feet higher than the floor level in the other two sections, while the upper floor was at the same street level as the center section. There was a sub-basement under the north wing.

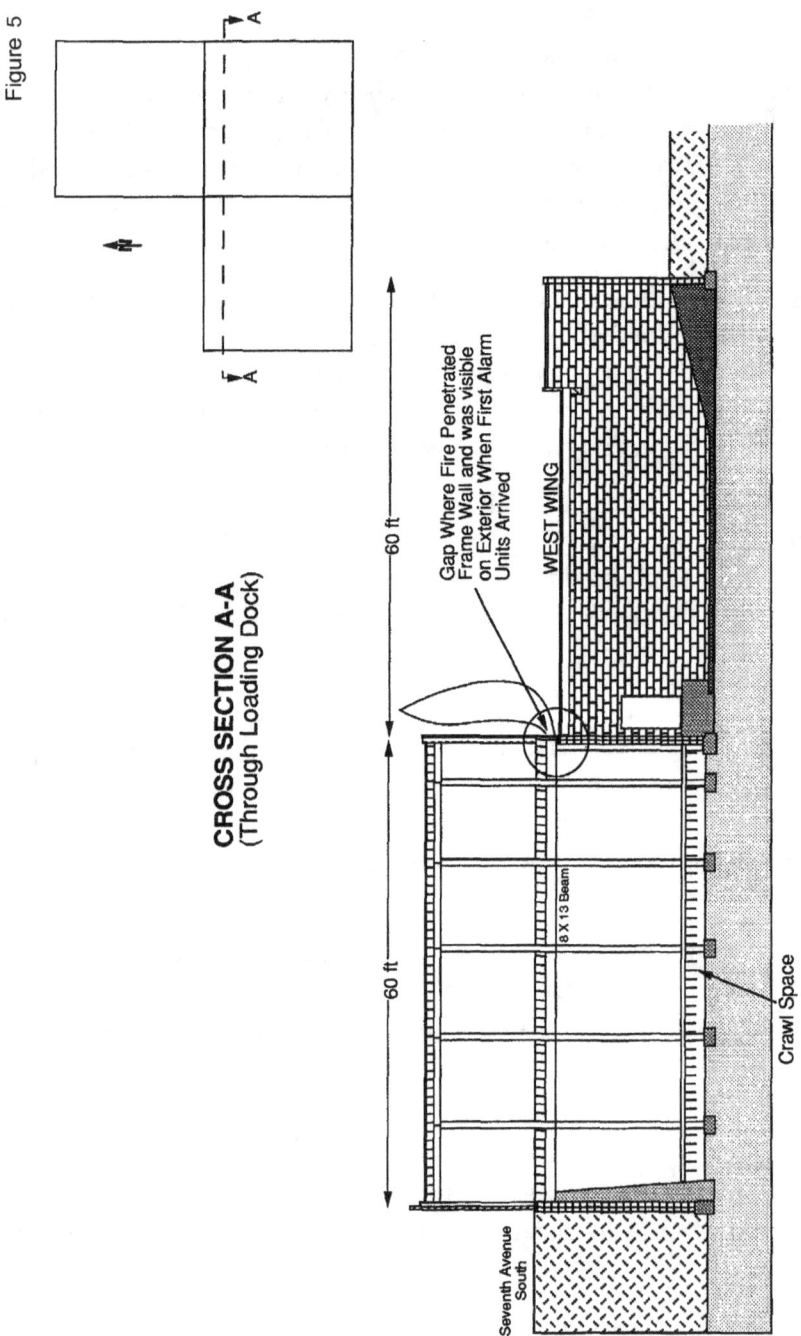

Figure 5

CROSS SECTION A-A
(Through Loading Dock)

Gap Where Fire Penetrated Frame Wall and was visible on Exterior When First Alarm Units Arrived

WEST WING

8 X 13 Beam

Crawl Space

Seventh Avenue South

Appendix A (continued)

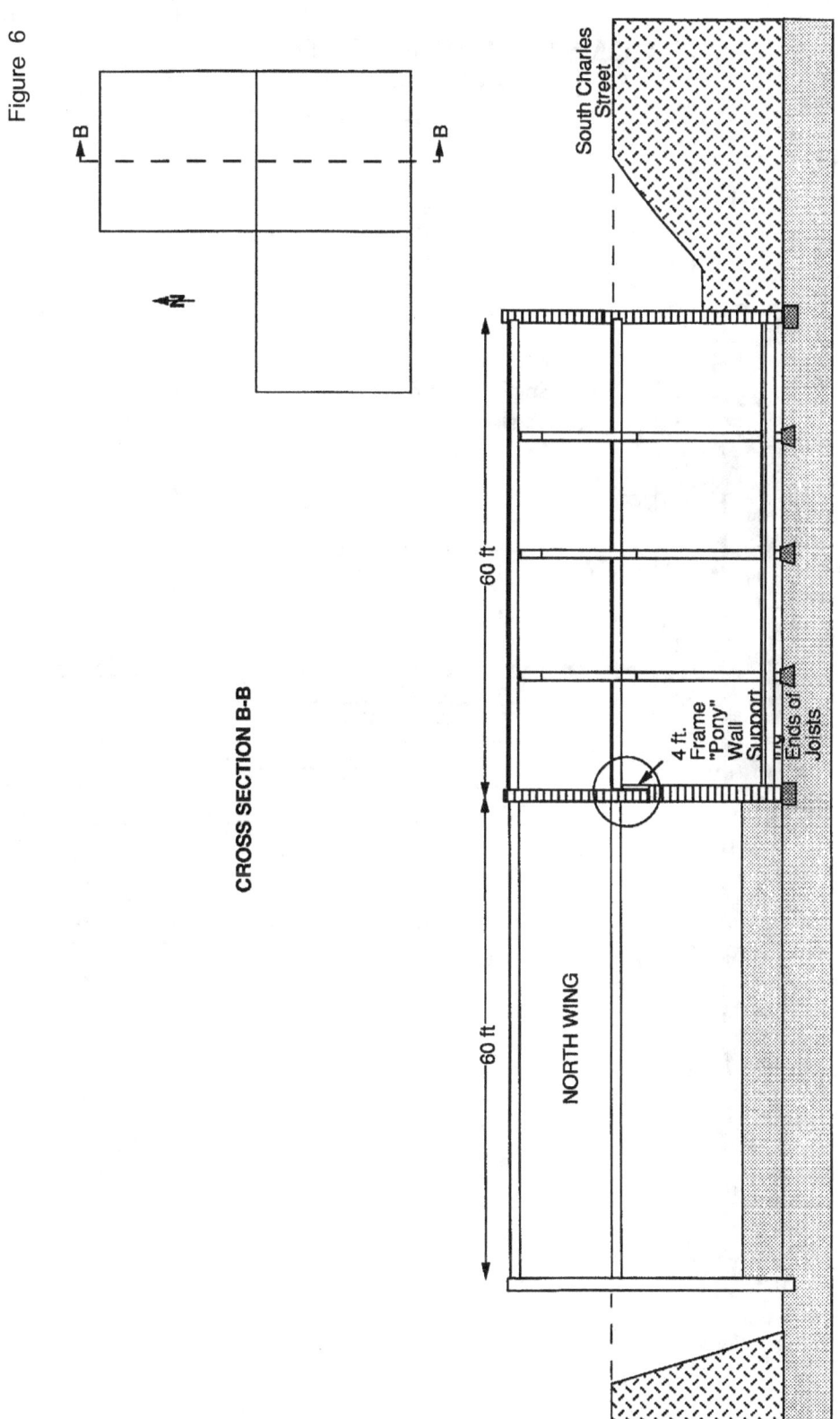

Figure 6

CROSS SECTION B-B

Appendix A (continued)

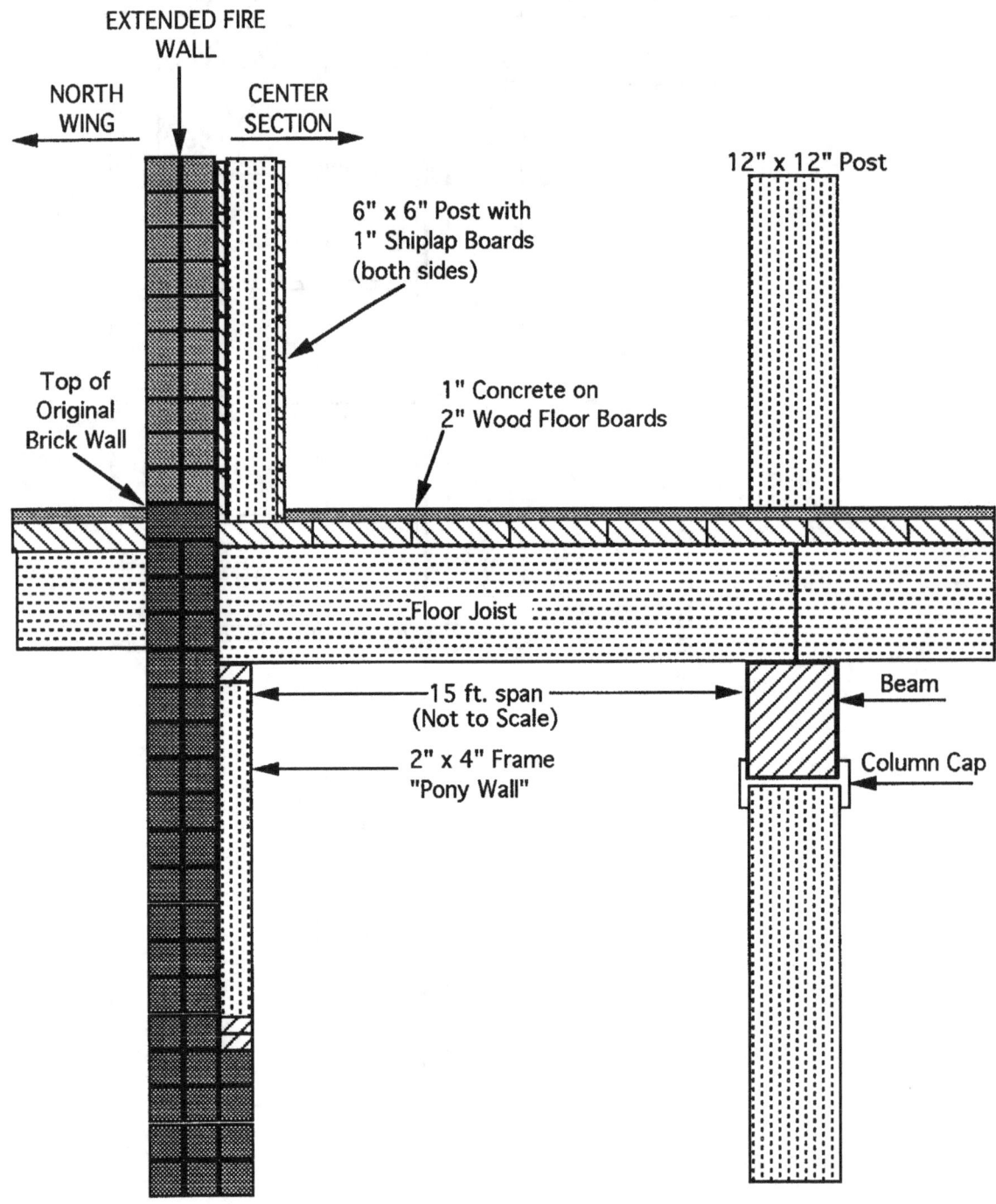

Figure 7

**CONSTRUCTION   DETAILS
AT INTERIOR FIRE WALL**
(ALL DIMENSIONS ARE ESTIMATES)

EXTENDED FIRE
WALL

NORTH
WING

CENTER
SECTION

12" x 12" Post

6" x 6" Post with
1" Shiplap Boards
(both sides)

Top of
Original
Brick Wall

1" Concrete on
2" Wood Floor Boards

Floor Joist

15 ft. span
(Not to Scale)

Beam

2" x 4" Frame
"Pony Wall"

Column Cap

Appendix A (continued)

Figure 8

FLOOR COLLAPSE MECHANISM
UPON FAILURE OF PONY WALL

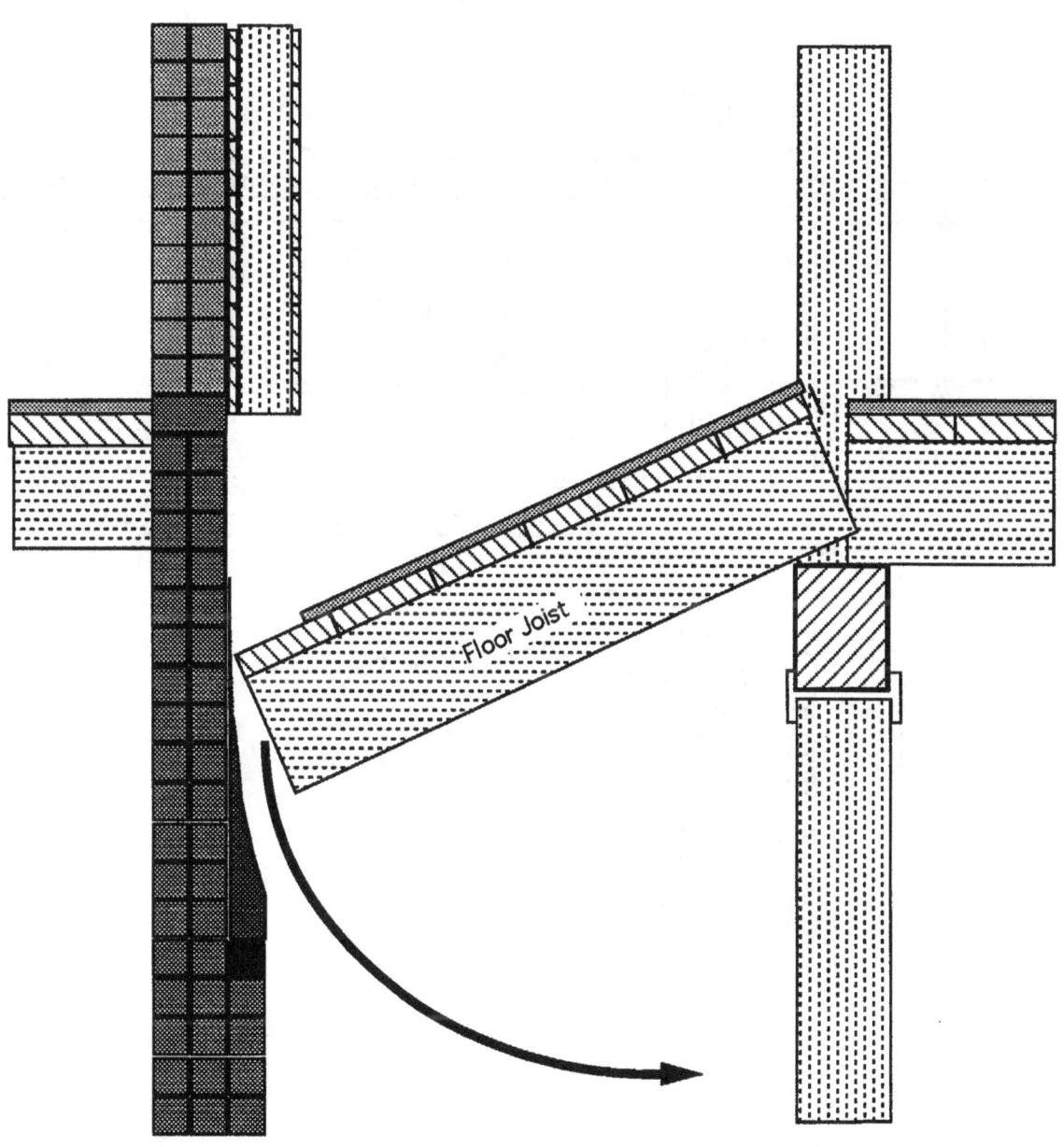

Appendix A (continued)

Figure 9

DETAIL OF UPPER FLOOR DECK
AND COLUMN CONNECTIONS
(All Dimensions are Estimated)

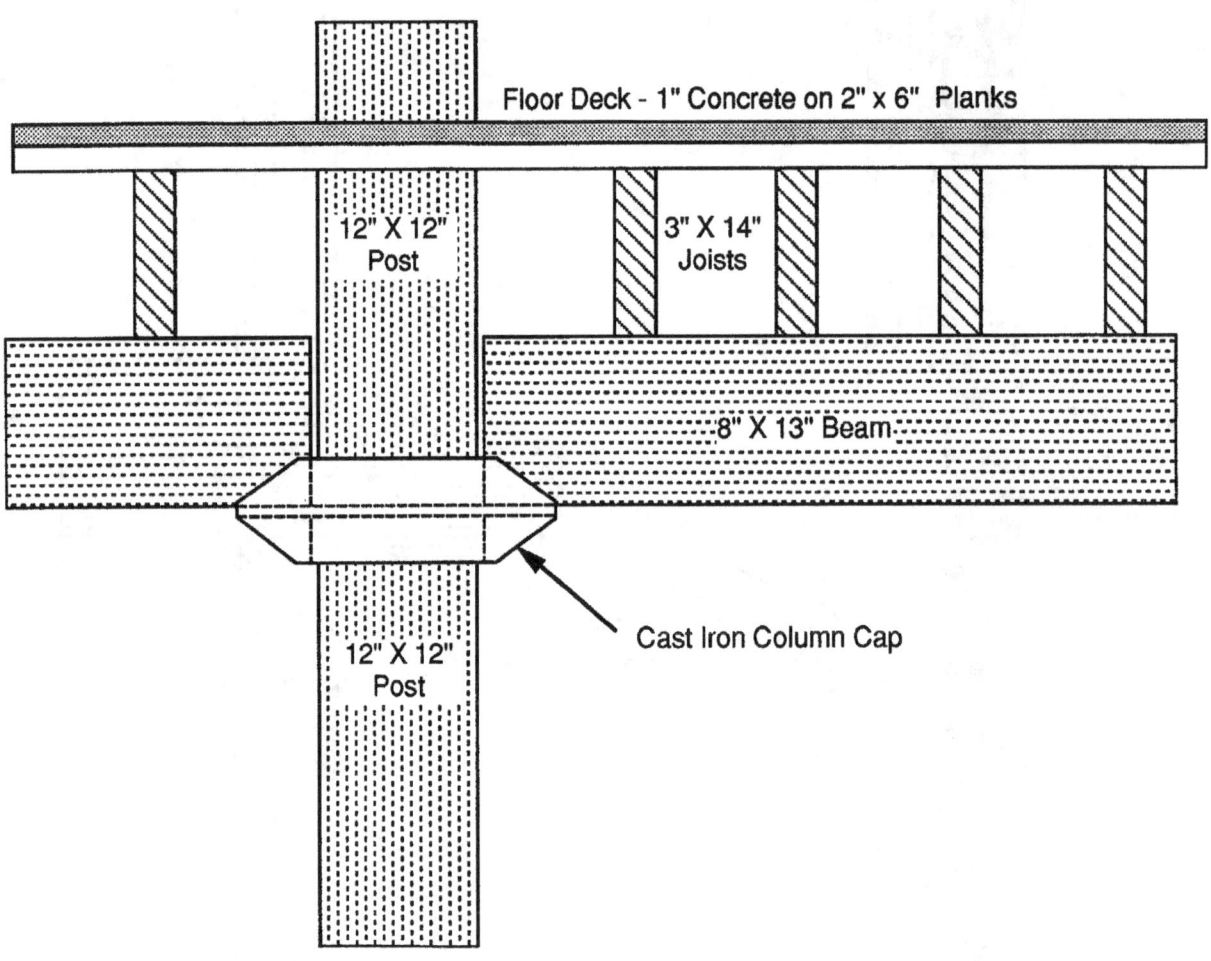

APPENDIX B

Photographs

View from the rear (west side) of the Mary Pang building, looking from South Charles Street. The open door is the entrance to the bakery. The heavy vegetation between South Charles Street and the building can be seen on the right side of the picture.

Appendix B (continued)

A large volume of fire was visible to units approaching from the west in the early stages of the incident. The outline of the lunchroom is visible next to the tallest flames.

Appendix B (continued)

Crews prepare to enter the structure on the east side.

Appendix B (continued)

A large volume of fire was visible to units approaching from the west in the early stages of the incident. The outline of the lunchroom is visible next to the tallest flames.

www.ingramcontent.com/pod-product-compliance
Lightning Source LLC
Chambersburg PA
CBHW081357170526

45166CB00010B/3125